SERVE ANALYSIS

IN LVF (italian women leagua A1) & SL (italian men leagua A1)

SEASON 2021-22

MATCHES IN THE WOMEN'S VOLLEYBALL LEAGUE 2021-22

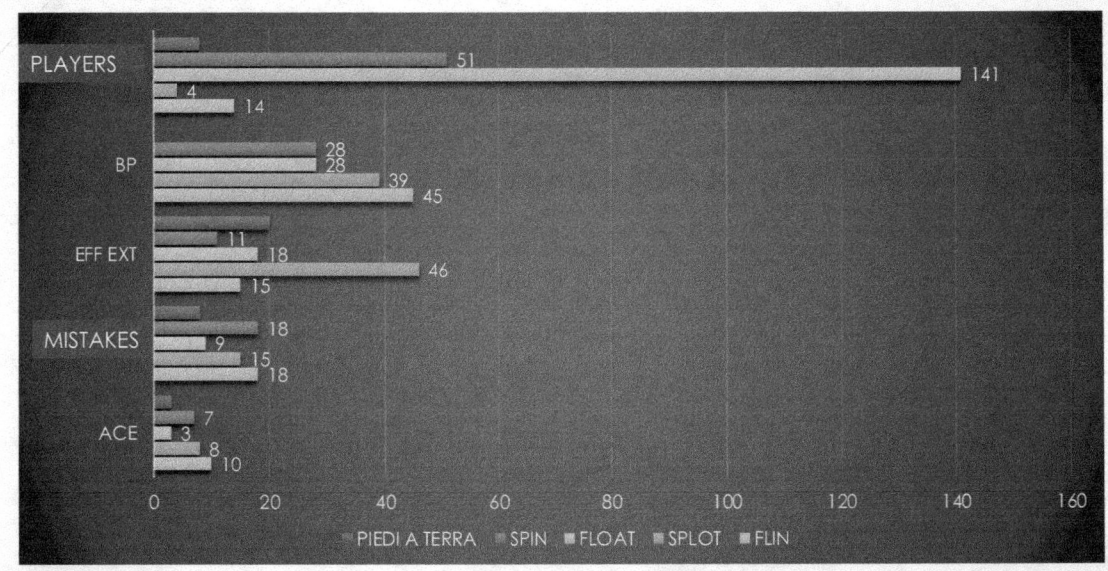

HOW AND WHO SERVE IN LVF 2021-22

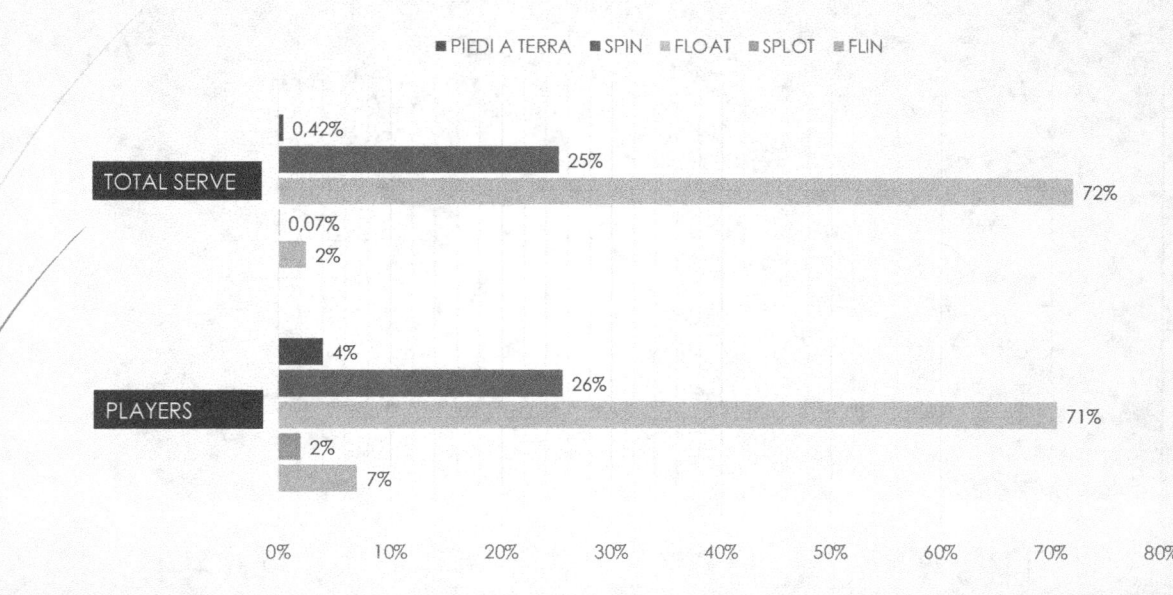

SERVE IN THE 2021-22 SUPERLEAGUE

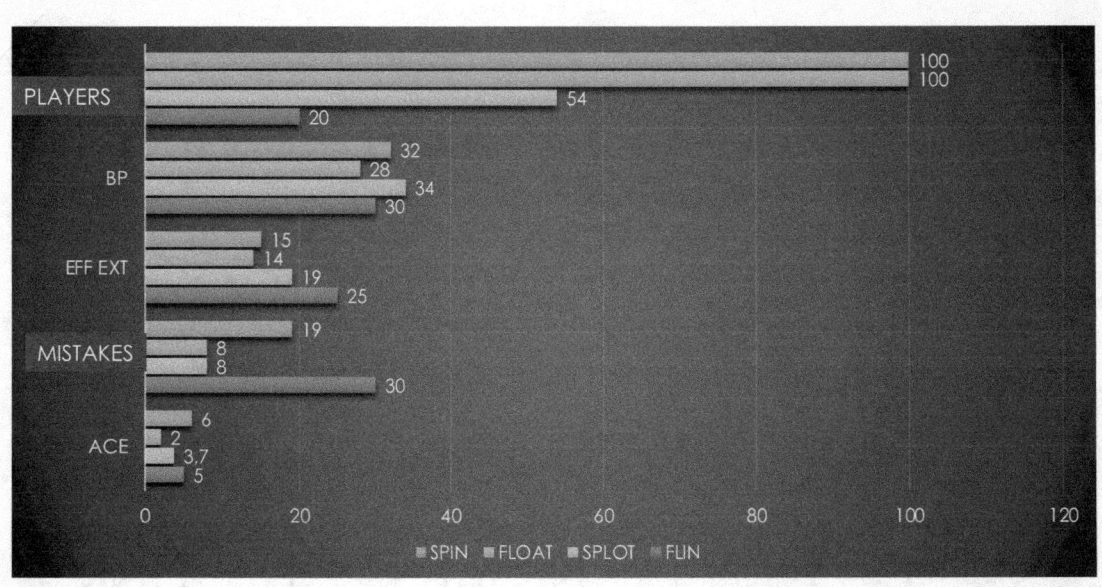

HOW AND WHO SERVE IN SL 2021-22

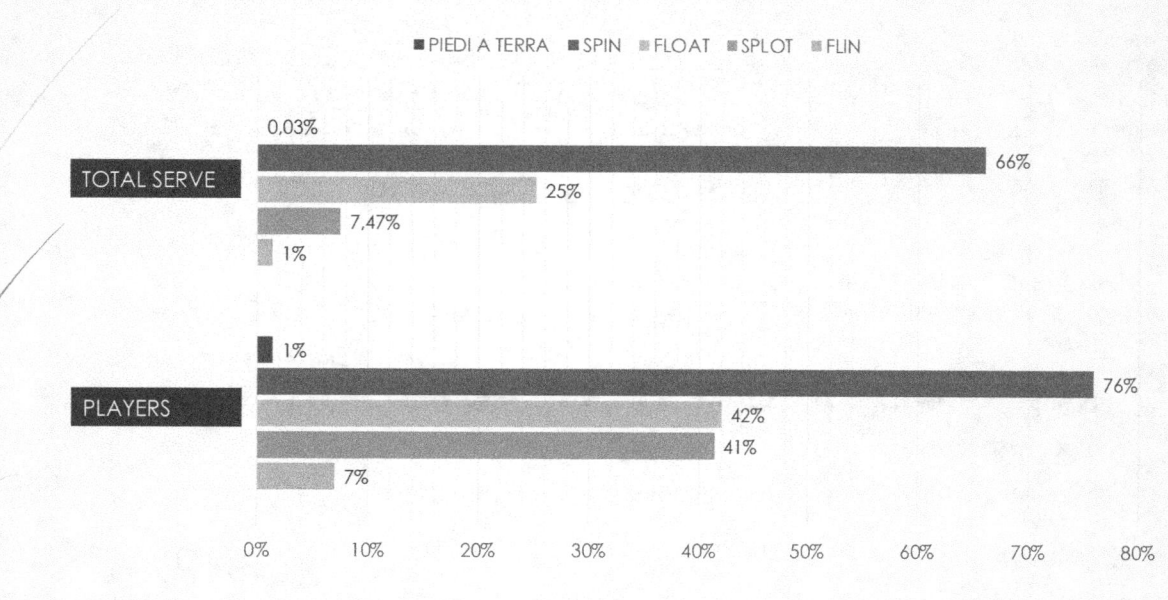

■ PIEDI A TERRA ■ SPIN ■ FLOAT ■ SPLOT ■ FLIN

HYBRID?

- DEFINITELY MIXED THEM THEY HAVE HIGHER % BP AND EXTENDED EFFEVEN THE TRADITIONAL JOKES

- HOWEVER, BY COMBINING DIFFERENT TECHNIQUES THEY ENCOUNTER EFFICIENCY PROBLEMS IN A SHORT TIME, WHICH RESULTS IN A HIGH ERROR %YOU MUST INITIALLY ACCEPT

- WITH RESPECT TO THE FEMALE HOWEVER, WHERE ONLY7% OF LVF FEMALE PLAYERS IN THE 2021-22 SEASON HE EXPLORED NEW TECHNIQUES IN RACING, IN SUPERLEAGUE 42% OF PLAYERS HAS EXPERIENCED AT LEAST ONE VARIATION, ALTHOUGH IN TOTAL IT APPEARS TO ONLY BE LESS THAN 9% OF THE TOTAL SERVICES.

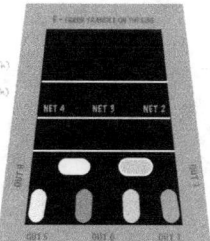

- Conflict 5/6, Speed strong *(25≥100 km/s OF>70 km/s)*
- Conflict 6/1, Speed strong *(35≥100 km/s OF>70 km/h)*
- On the line on 6, Speed *(35>90 km/h OF>62 km/s)*
- On the line on 1, Speed *(35>90 km/h OF>44 km/h)*
- Short Drop, Speed *(25 and OF between 56/55 km/s)*

STRONG target: *include only strong serve (R-7-Q-8)*

SHORT target: *include only short & drop serve (W-W/D)*

NET 4   NET 5   NET 2

OUT 5   OUT 6   OUT 1

Dk Overhead

UpRight   UpLeft

Right   Body   Left

LowRight   LowLeft

Low

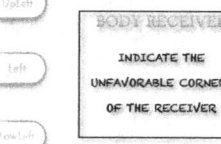

BODY RECEIVER

INDICATE THE UNFAVORABLE CORNERS OF THE RECEIVER

LEGEND

JF SERVE JUMP FLOAT
JS SERVE JUMP SPIN
FLIN SERVE WITH TOSS FLOAT , SHOT SPIN
SPLOT SERVE WITH TOSS SPIN , SHOT FLOAT
BP SHOW HOW MUCH BREAK POINT
BP ON TARGET WITHIN THE SERVE ON TARGET
HIGH JUMP SHOW BEST JUMP
AVERAGE JUMP SHOW AVERAGE JUMP
CHALLENGE ASSING A POINT FOR EACH DAY YOU EXCEED YOUR GOAL

EFFICIENCY THE RESULT OF THIS FORMULA (#/)-(=)
EXTENDED EFFICIENCY THE RESULT OF THIS FORMULA (#/+)-(=)
QUOTE EFFICIENCY THE RESULT OF THIS FORMULA (#/+!)-(=)

EVALUATION EFFECT
\# Ace
\+ Pass from net
\- Pass near net
/ Balls return without opponent spike
! Pass on the 3 mt. Line
= Mistake

SERVE ON THE PERIMETER

- IT APPEARS CLEAR HOW TO SEARCH THE TARGETS OF THE ROWS OR IN CONFLICT 5/6 OFFER ONE% BP VERY HIGH (+30%)

- BEATING THE PLAYER OR SEEKING TARGETS OFFERS A % OF EFF ALSO EXTENDED BY +60%

- WHAT A BEAT UP NOT TARGET IT HAS THE SAME PROBABILITY OF BP AS ONE SHORT JOB BUT THE SHORT HAS A VERY HIGH EXTENDED EFFICIENCY %.

- THE PERIMETER AND THE SEARCH OF LINE 1 (B) SHOULD BE MORE DARED.

HOWEVER LESS THAN 30%-39% OF SERVICES END IN TARGET OR CONFLICTS

THE REAL QUESTION IS:

- IN A SPORT WHERE YOU ARE ASKED TO ATTACK A BALL IN FLIGHT AT 100KMH AND TO HIT A MOVING PHALANX

- IN A SPORT WHERE EVERYTHING IS ANALYZED, JUMPS, HEIGHT, POWER, EFFECTS, PHYSICAL STATE ETC...

- IN A SPORT WHERE THOUSANDS OF REPETITIONS OF A GESTURE ARE PERFORMED

- REALLY IN THE ONLY FUNDAMENTAL «HALF CLOSE SKILL» THAT IS INFINITELY REPEATABLE CAN YOU NOT ASK FOR A DEGREE OF PRECISION LESS THAN 3 METERS OF FIELD?

»I THINK THAT IN 10 YEARS THE REQUIREMENT TO BIT TO THE RIGHT OR LEFT, ON THE KNEE OR THE SHOULDER, ON THE LINE RATHER THAN A FALLING BALL, WILL BECOME THE NORMALITY EVEN AT 100KMH. WHAT IS MISSING NOW IS THE CONVICTION THAT YOU CAN ASK AND GET IT."

Mauro Marchetti

TARGET ON THE NET

- THE BEAT THAT AFFECTS THE TAPE HAS VERY HIGH % BP AND EXTENDED EFF

- BUT TRAINING TO HIT THE RIBBON?

THE »HOLY» SHORT JOKE

- VERY LOW ERROR %.

- HIGH ACE AND SLASH % FOR FREQUENT RECEIVING THE LEAST SPECIALIST

- % OF EFF EXTENDED VERY HIGH

- JOKE THAT REMOVES 25-50% OF NETWORK POINTS TO BE CONSIDERED

- VERY HIGH BREAK POINT %.THANKS TO THE MURO DEFESA ORGANIZATION

- JOKE THAT REQUIRES ONE BASIC TECHNIQUE OF RECEPTION

- QUOTE THAT HIGHLIGHTS PHYSICAL PROBLEMS OR DEFICIENCIES - TIBIO TARSIAL OR LOWER LIMBS STRENGTH -

- TECHNIQUE TO BE LEARNED VERY QUICKLY AND GIVEN THE SPEED AND ANGLES OBTAINABLE EVEN UNDER STRESS

TYPES OF MISTAKES IN JUMP FLOAT LVF

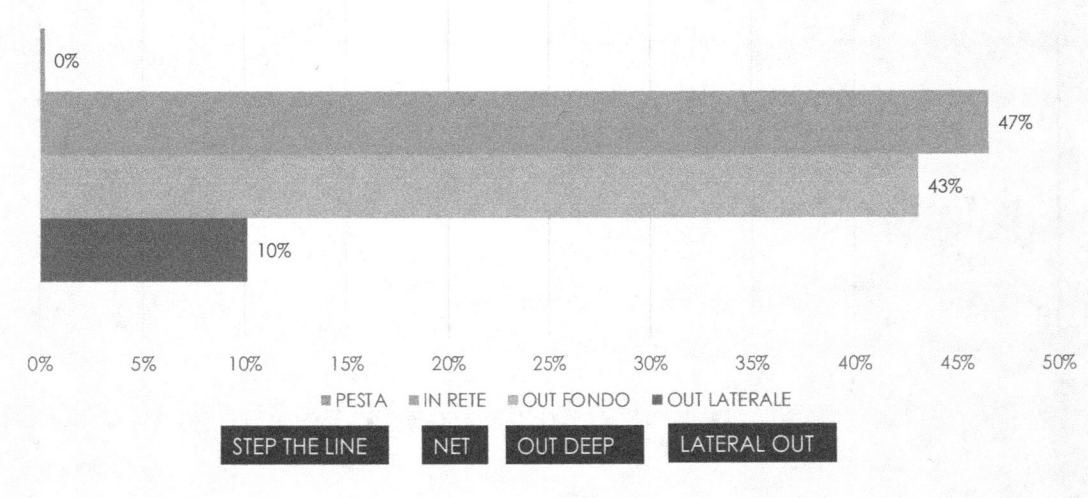

0%

47%

43%

10%

0% 5% 10% 15% 20% 25% 30% 35% 40% 45% 50%

■ PESTA ■ IN RETE ■ OUT FONDO ■ OUT LATERALE

STEP THE LINE   NET   OUT DEEP   LATERAL OUT

TYPES OF MISTAKES IN JUMP SPIN SL

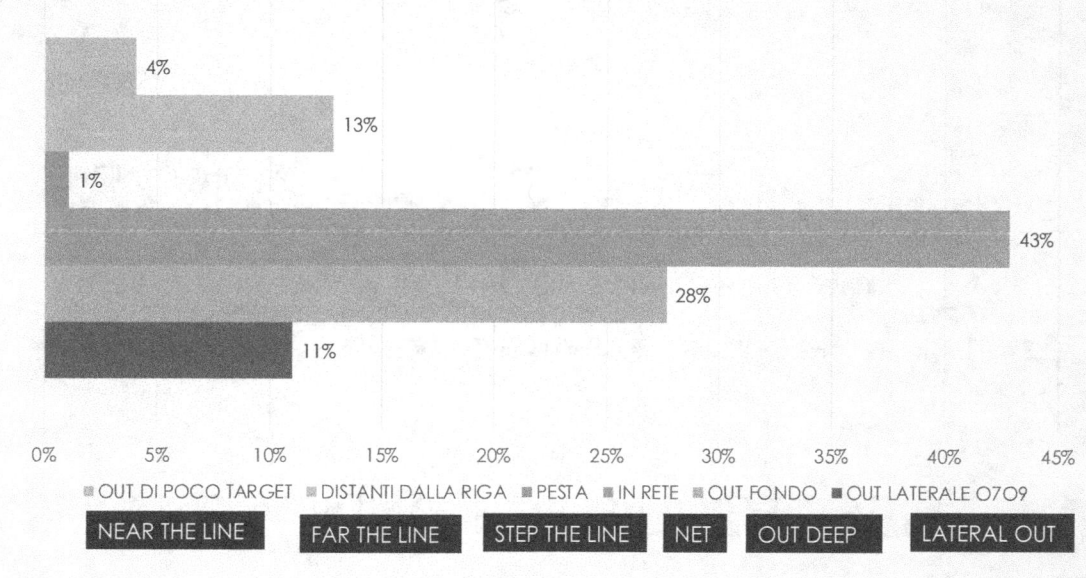

4%

13%

1%

43%

28%

11%

0%   5%   10%   15%   20%   25%   30%   35%   40%   45%

■ OUT DI POCO TARGET ■ DISTANTI DALLA RIGA ■ PESTA ■ IN RETE ■ OUT FONDO ■ OUT LATERALE 0709

| NEAR THE LINE | FAR THE LINE | STEP THE LINE | NET | OUT DEEP | LATERAL OUT |

SERVE LOOKING FOR SIDE TARGETS

- PRECISELY GIVEN THE GREAT ADVANTAGE OF BEATING ON THE SIDE LINES IT IS NOT EXPLAINED THE SMALL ERROR PERCENTAGE - 10% - ON BALLS THAT COME OUT LATERALLY(I.E. ON THE LONG SIDE OF THE FIELD).

- IF IT IS TRUE THAT MOST COACHES PREFER THE ERROR OUTSIDE THE BALL INTO THE NET, THE NUMBERS OFTEN TELL US THAT THE REALITY IS DIFFERENT.DO WE MONITORING IT?

THE QUESTION IS' :

HOW MUCH DO THEY LISTEN TO YOU?

HOW MUCH DO THEY KNOW WHAT THEY DO?

HOW MUCH DO THEY DO WHAT THEY WANT?

HOW MUCH DO THEY WANT TO DO WHAT YOU WANT?

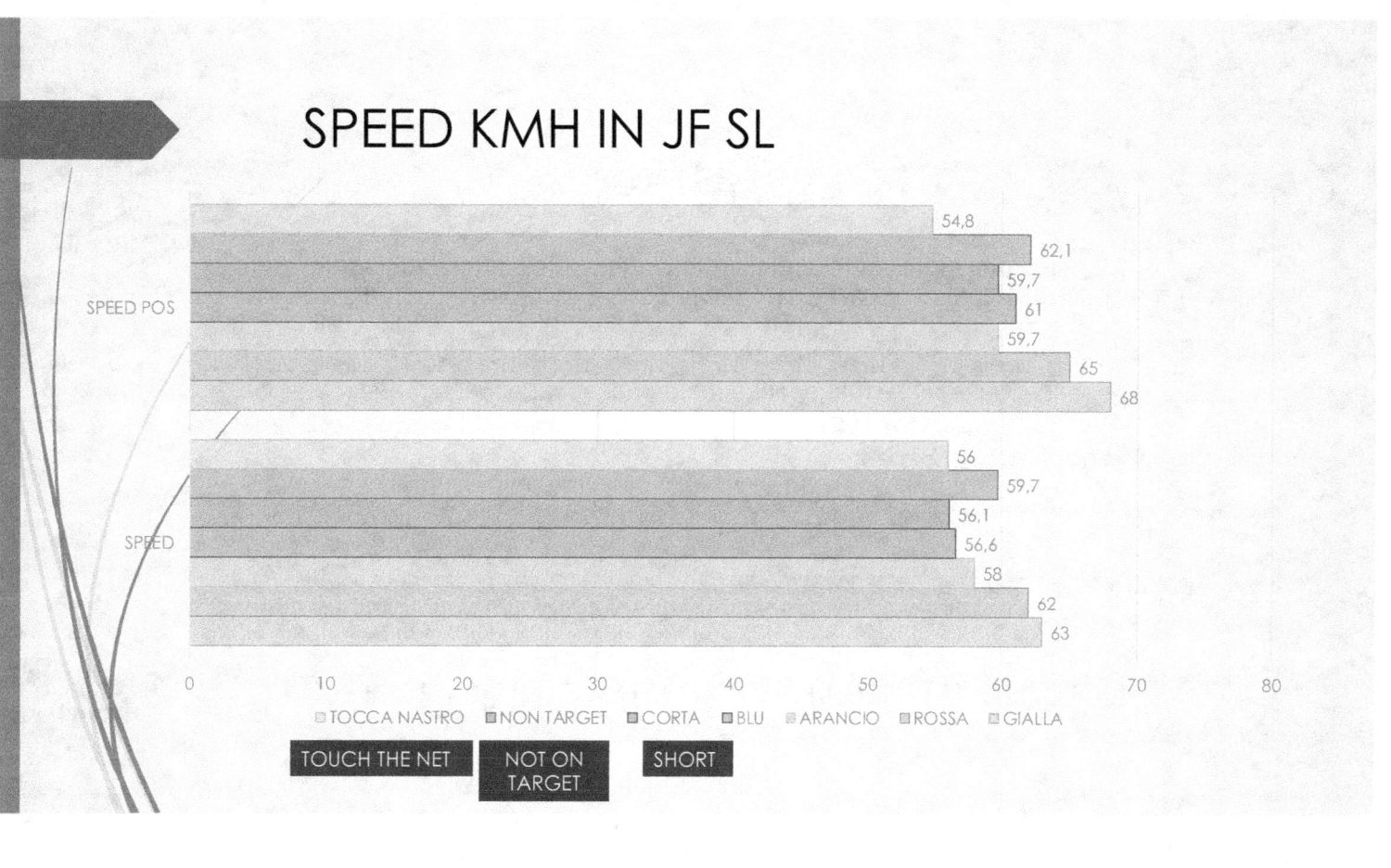

SPEED KMH IN JF SL

SPEED AND REQUIREMENTS LF SL

- THE AVERAGE SPEED IS BETWEEN 59-65 KMH

- IT APPEARS CLEAR THAT ON AVERAGE THE POSITIVE CALL TRAVELS 3-5 KMH MORE THAN THE NEGATIVE CALL (RECEIVED - O!)

- EVEN THE SHORT BALL, ALTHOUGH ITS IDEAL SPEED FROM AN AERODYNAMIC POINT OF VIEW IS 44KMH, YOU OBTAIN A POSITIVE MIX IN THE BALANCE OF SPEED AND NET PASSING AT 54KMH. IN ALL CASES THE LOWER THE SPEED THE BETTER THE EXTENDED EFFICIENCY.

- THE JOKE ABOUT YELLOW TARGET IT GETS AN AVERAGE SPEED OF 5KMH HIGHER – DUE TO THE FACT THAT THE FLOAT IS MAINLY CARRIED OUT BY THE ZONE 5 STATIONS.

- THE BALL THAT GET POSITIVE EFFECTS BY TOUCHING THE TAPE IS VERY SLOW, 54-6KMH, AND GET MORE BENEFITS THE SLOWER IT IS, HOWEVER MAKING IT DIFFICULT TO TRAIN IT BECAUSE IT IS A BALL WITH A PARABLE AND ONLY ONE PASS TO THE NET.

YOU CAN AND SHOULD KNOW THE SPEED AND ASK TO BEAT HARDER

SPEED KMH IN JF LVF

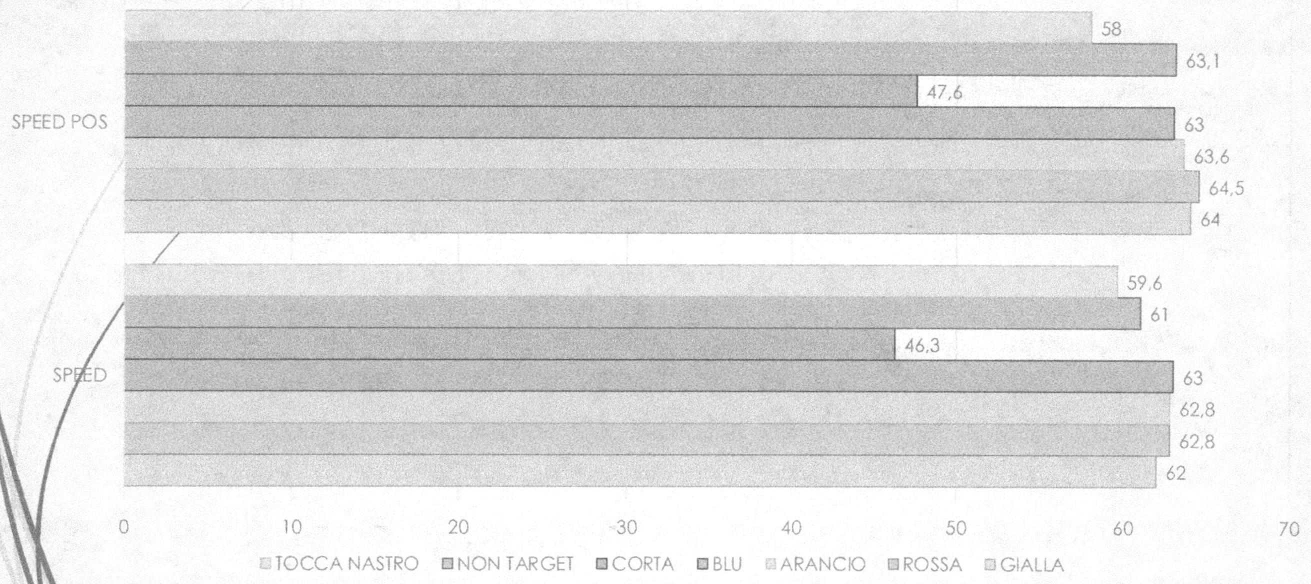

SPEED POS
- 58
- 63,1
- 47,6
- 63
- 63,6
- 64,5
- 64

SPEED
- 59,6
- 61
- 46,3
- 63
- 62,8
- 62,8
- 62

| 0 | 10 | 20 | 30 | 40 | 50 | 60 | 70 |

■ TOCCA NASTRO ■ NON TARGET ■ CORTA ■ BLU ■ ARANCIO ■ ROSSA ■ GIALLA

SPEED AND REQUIREMENTS JF LVF

- THE AVERAGE SPEED IS BETWEEN 61-64 KMH
- IT APPEARS CLEAR THAT ON AVERAGE THE POSITIVE CALL TRAVELS 2 KMH MORE THAN THE NEGATIVE CALL (RECEIVED - O!)
- THE CENTRALS PULL ON AVERAGE UNDER 58KMH
- EVEN THE SHORT BALL, ALTHOUGH ITS IDEAL SPEED FROM AN AERODYNAMIC POINT OF VIEW IS 44KMH, YOU OBTAIN A POSITIVE MIX IN THE BALANCE OF SPEED AND NETWORK PASSAGE AT 47KMH.
- THE BALL THAT OBTAINS POSITIVE EFFECTS BY TOUCHING THE TAPE IS VERY SLOW, 58KMH, AND OBTAINS MORE BENEFITS AS MORE 'IT IS SLOW, HOWEVER MAKING IT DIFFICULT TO TRAIN IT BECAUSE IT IS A BALL WITH A PARABLE AND ONLY ONE PASS TO THE NET.

YOU CAN AND SHOULD KNOW THE SPEED AND ASK TO SERVE HARDER

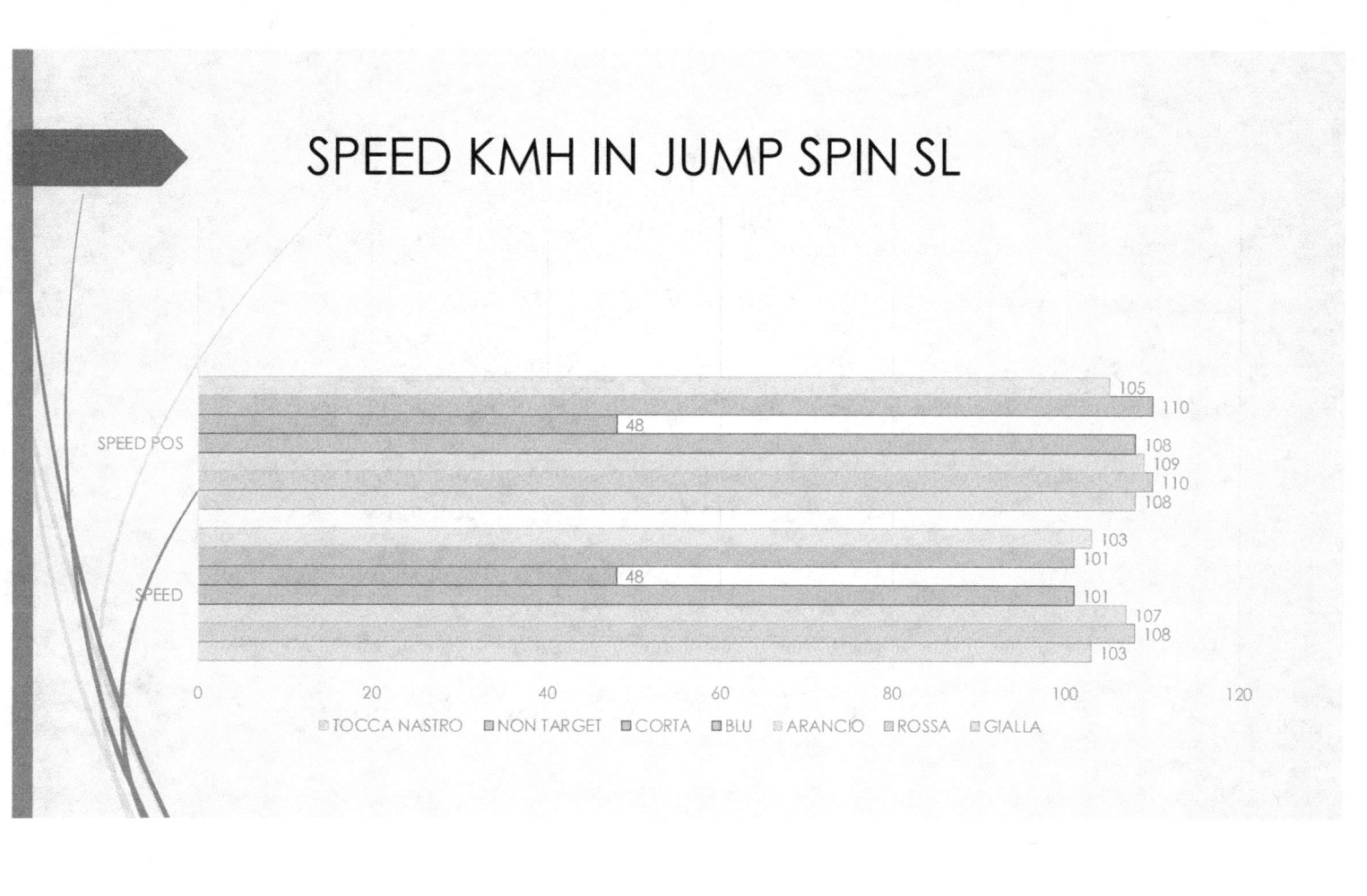

SPEED AND REQUESTS JS SL

- THE AVERAGE SPEED IS BETWEEN 101-103 KMH

- IT APPEARS CLEAR THAT ON AVERAGE THE JOB IN CONFLICTS BECOMES POSITIVE IF IT TRAVELS 7-10 KMH MOREOF THE NEGATIVE ONE (RECEIVED - O!)

- THE BIT IN THE BLUE AND YELLOW LINES REQUIRE NOT ONLY THE TARGET BUT A HIGH SPEED - AROUND 108KMH - TO BE CHALLENGING AND OBTAIN BENEFITS.

- ALONE ABOVE 115KMH THE TARGET THEY ARE BEGINNING TO BE NOT INDISPENSABLE

YOU CAN AND SHOULD KNOW THE SPEED AND ASK TO SERVE HARDER

SPEED KMH IN JS LVF

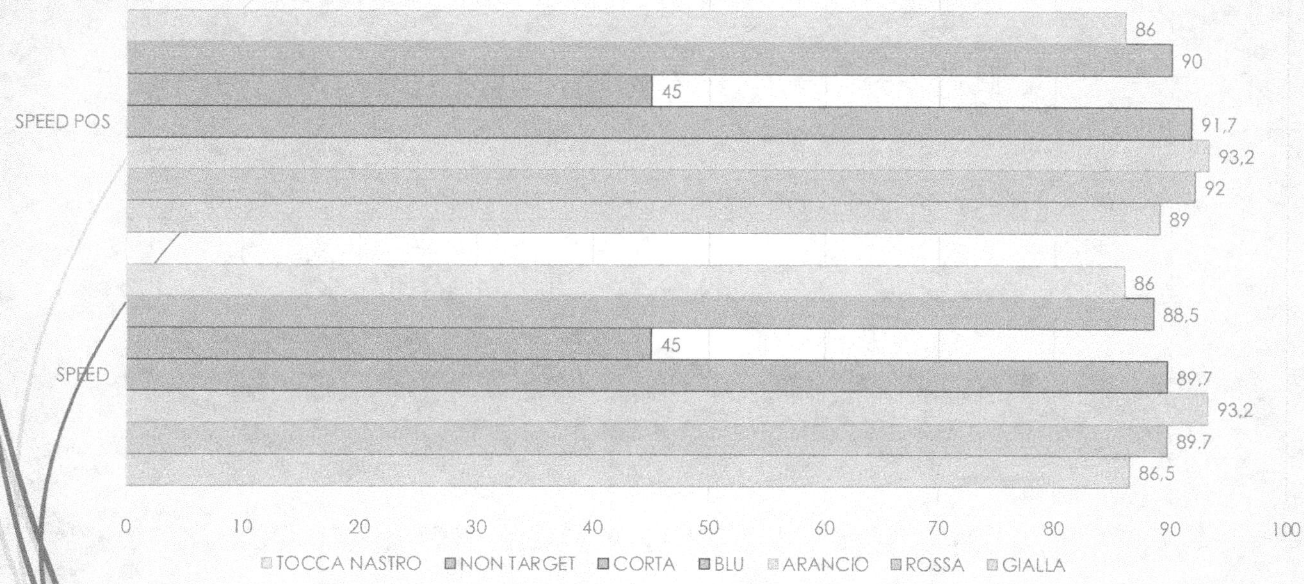

Bar chart with two groups, **SPEED POS** and **SPEED**, plotted on a horizontal axis from 0 to 100.

SPEED POS:
- 86
- 90
- 45
- 91,7
- 93,2
- 92
- 89

SPEED:
- 86
- 88,5
- 45
- 89,7
- 93,2
- 89,7
- 86,5

Legend: ☐ TOCCA NASTRO ☐ NON TARGET ☐ CORTA ☐ BLU ☐ ARANCIO ☐ ROSSA ☐ GIALLA

JS LVF SPEED AND REQUESTS

- THE AVERAGE SPEED IS BETWEEN 93-95 KMH

- IT APPEARS CLEAR THAT ON AVERAGE THE JOB IN CONFLICTS BECOMES POSITIVE IF IT TRAVELS 2-3 KMH MORE THAN THE NEGATIVE ONE (RECEIVED - OR!)

- A JS A JOKE90KMH IS A BACKGROUND PLACE UNLESS IT HAS THE CHANCE TO SURPRISE

YOU CAN AND SHOULD KNOW THE SPEED AND ASK TO SERVE HARDER

SPEED AND MISTAKES JF

- YOU SHOOT AT THE NET BECAUSE YOU SHOOT TOO SLOWLY
- IT PULLS OUT BECAUSE IT PULLS TOO HARD

- THE TRUTH IS THAT THE STRONG AND TENSIVE BIT AND THE SLOWLY AND PARABLE ONE HAVE DIFFERENT BODY-SPACE-BALL RELATIONSHIPS
- A PLAYER USED TO SHOT HARD WHO IS ASKED TO MANAGE IT OFTEN ONLY REMOVES THE SPEED WITHOUT CHANGING THE RATIO AND THE BALL ENDS IN THE NET
- A PLAYER USED TO PLACING THE BALL WHICH IS ASKED TO FORCE, WILL ONLY INCREASE THE FORCE WITHOUT CHANGING THE RATIO AND THE BALL WILL END UP

«THE KNOWLEDGE OF THE TECHNIQUE IS LACKED AND THE AWARENESS OF WHAT IS BEING DONE IS LACKED».

SPEED AND MISTAKES JS SL

- YOU SHOOT AT THE NET BECAUSE YOU SHOOT TOO SLOWLY
- IT PULLS OUT BECAUSE IT PULLS TOO HARD

- THE TRUTH IS THAT THE BIT SPIN BELOW 90-95KMH LVF & 105-108KMH SLIT MAKES LITTLE SENSE, BUT THIS IMPLYES THAT WITH THIS SPEED TO GET ON THE COURT IT MUST HAVE A GLASS PASS, BE GRIPPED HIGH AND HAVE LITTLE SPIN (THE SPIN TAKES SPEED OUT OF THE BALL).
- SO 100 KMH ARE STANDARD SERVES WITH A LOT OF SPIN AND HIGH NET PASS - PLACED FROM THE BOTTOM OF THE COURT - WHICH BRING FEW ERRORS BUT ALSO FEW BENEFITS (32% BP IS THE LEAST % AMONG ALL THE TYPES OF SERVES).

«THE KNOWLEDGE OF THE TECHNIQUE IS LACKED AND THE AWARENESS OF WHAT IS BEING DONE, WHEN IT IS DONE AND WHY IT SHOULD BE DONE IS LACKED».

JS SL MISTAKES MOMENTS

ON 0-0 POINT BETTER PASS ALWAYS , STATISTIC 21% MISTAKES AND 71% FBSO

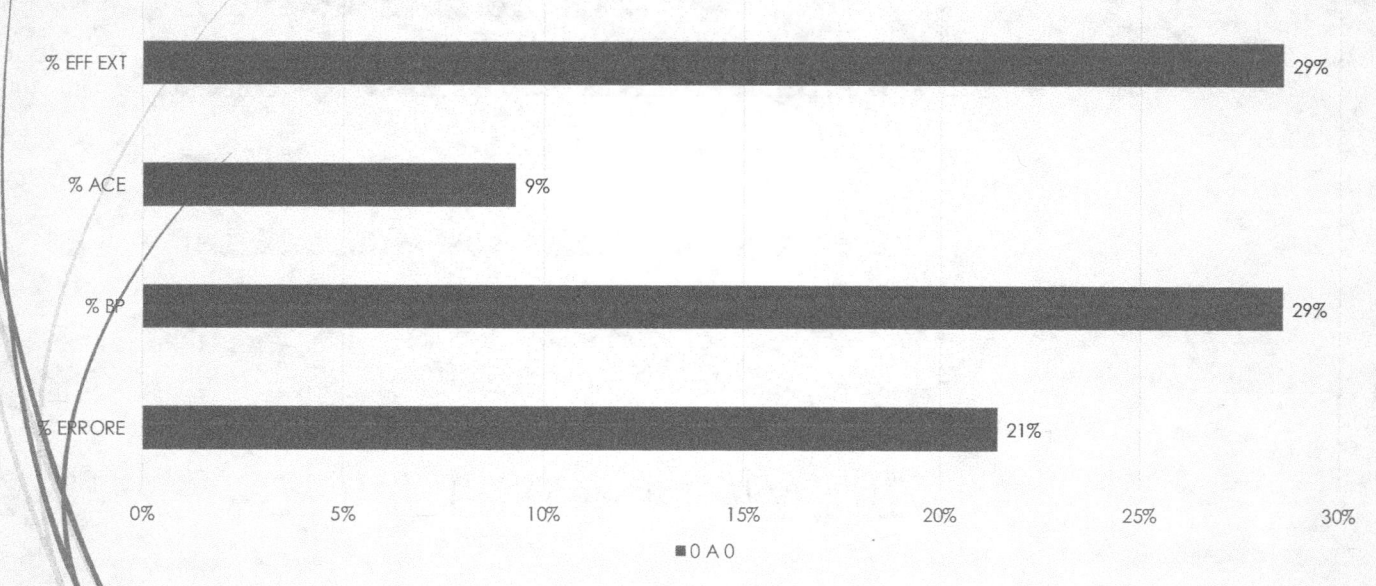

% EFF EXT	29%
% ACE	9%
% BP	29%
% ERRORE	21%

0%   5%   10%   15%   20%   25%   30%

■ 0 A 0

JS LVF ERROR MOMENTS

ON 0 A 0 ALWAYS BETTER PASSING , FBSO IS 63%

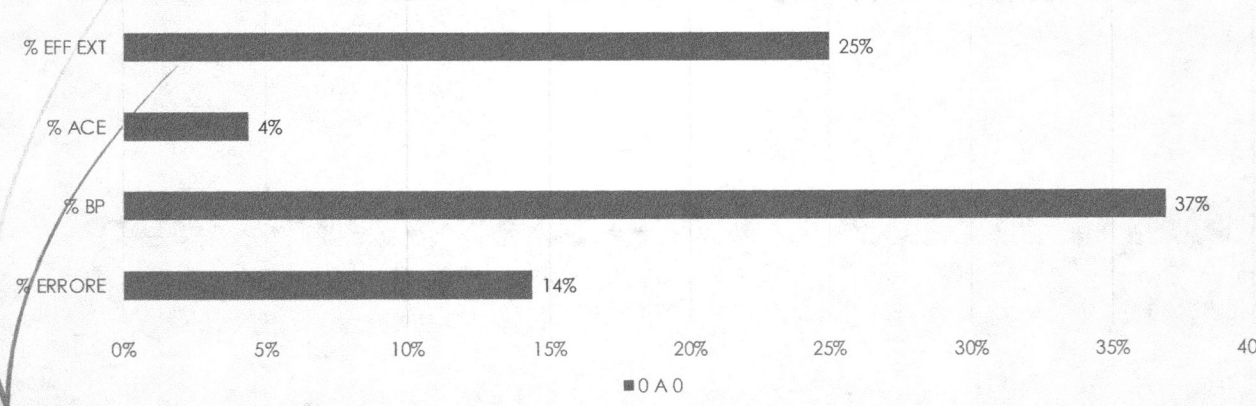

JF SL ERROR MOMENTS

ON 0-0 FBSP IS AT 81%

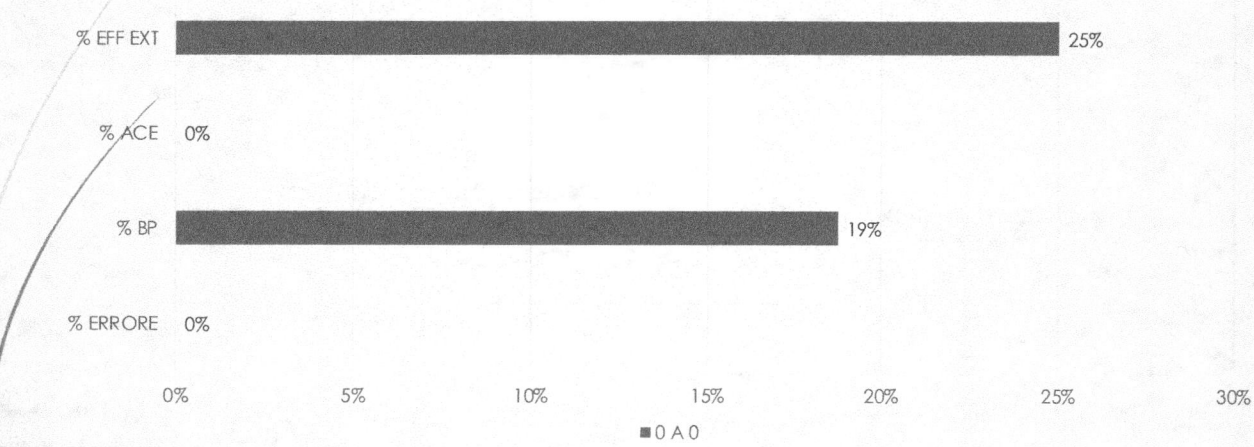

JF LVF ERROR MOMENTS

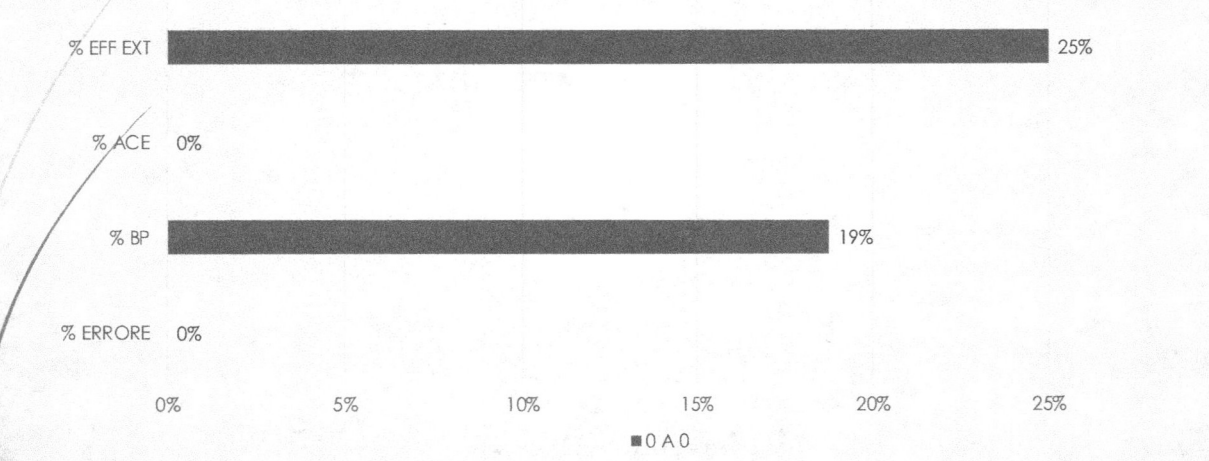

ON 0-0 BETTER PASSIN , FBSP IS AT 81%

% EFF EXT	25%
% ACE	0%
% BP	19%
% ERRORE	0%

0%   5%   10%   15%   20%   25%   30%

■ 0 A 0

RESULT OF JF SL INTERRUPTIONS

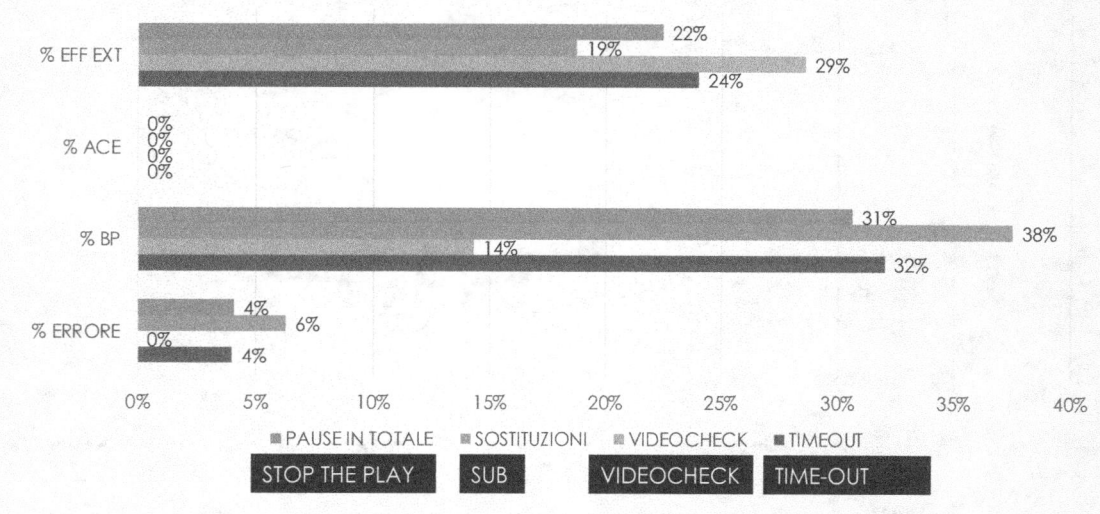

RESULT OF JS SL INTERRUPTIONS

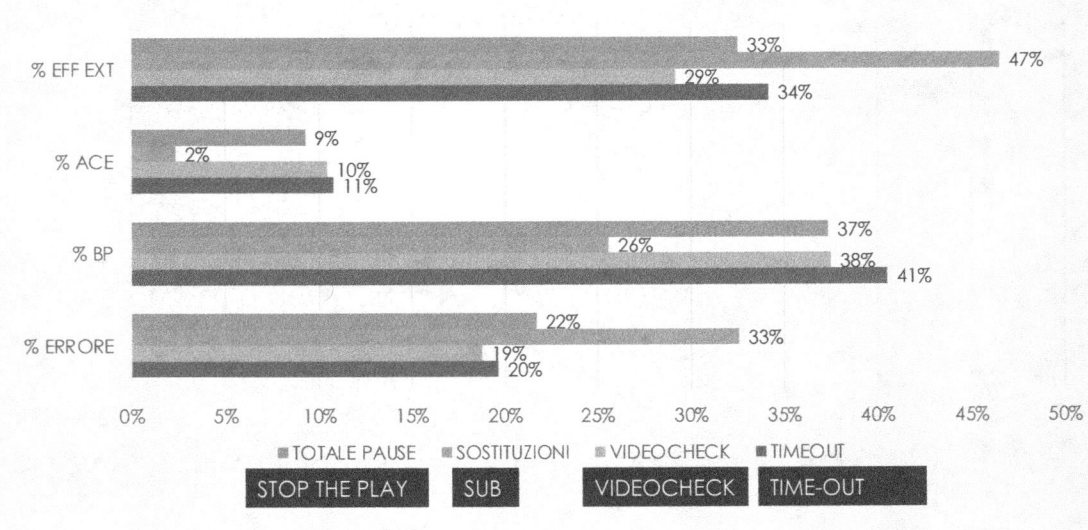

RESULT OF JF LVF INTERRUPTIONS

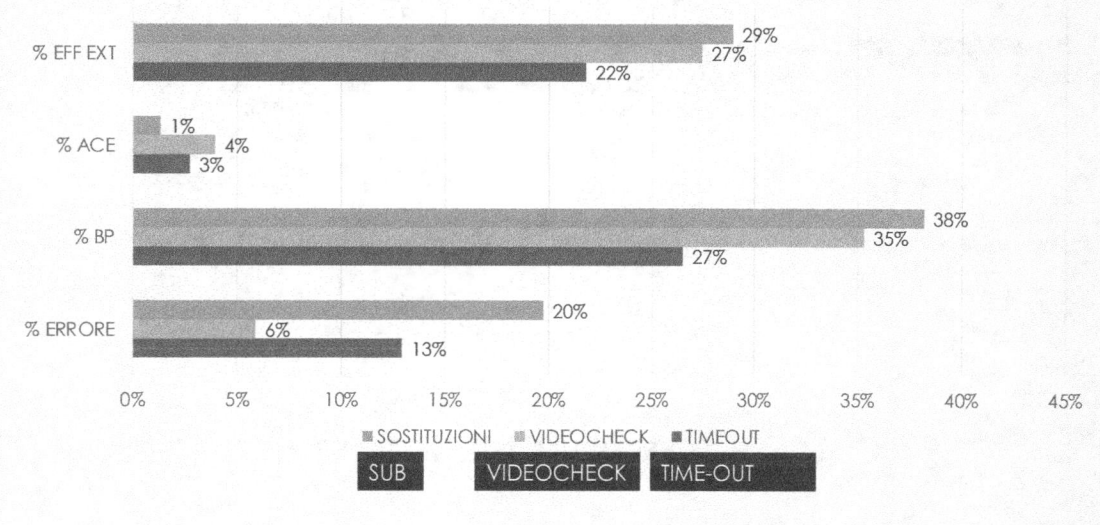

RESULT OF JS LVF INTERRUPTIONS

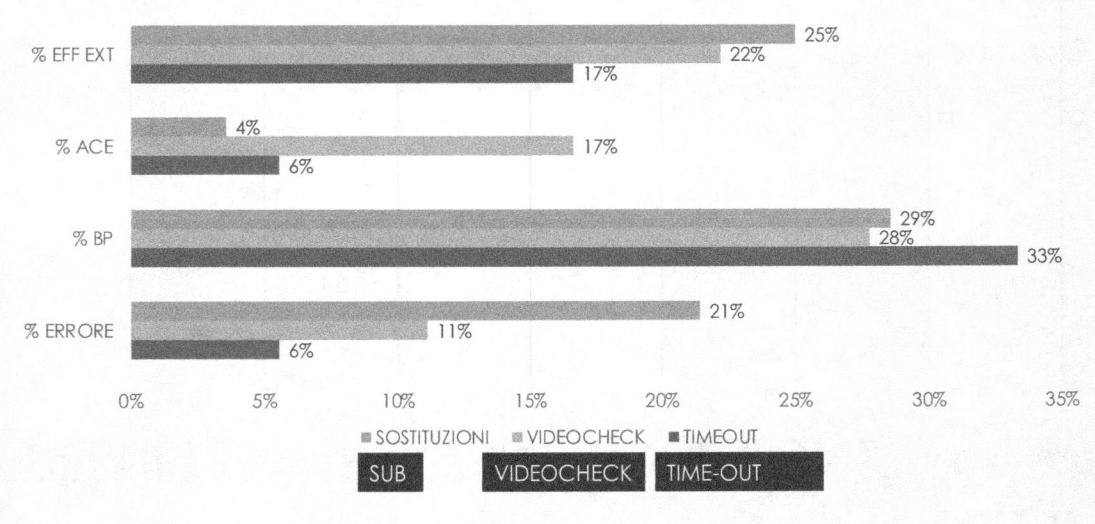

CALL TIME OUT TO BREAK?

- A TIME OUT AND A SUBSTITUTION DEFINITELY HAVE DESIRED EFFECTS IN THE ARRIVING BATTER:
- THERE ERROR % INCREASE
- EXTENDED EFFICIENCY DECREASES
- DECREASES DRASTICALLY THE % OFACE
- SO CALL TIME OUT - BETTER IF WITH THE ADVANTAGES (+10% ERR) - IT GUARANTEES AT LEAST TO MAKE OUR OPPONENT BEAT WORSE

- STRANGELYVIDEOCHECK DOES NOT HAVE THE SAME EFFECT, PERHAPS BECAUSE THE EXTENDED DURATION OF THE SUSPENSION CLEARS TENSIONS AND MAKES THE NEW SERVICE BECOME A MOMENT IN ITSELF AND NOT ONE LOADED WITH EMOTIONAL TENSION.

JOKE SERVE THAT FAL DOWN W & WA

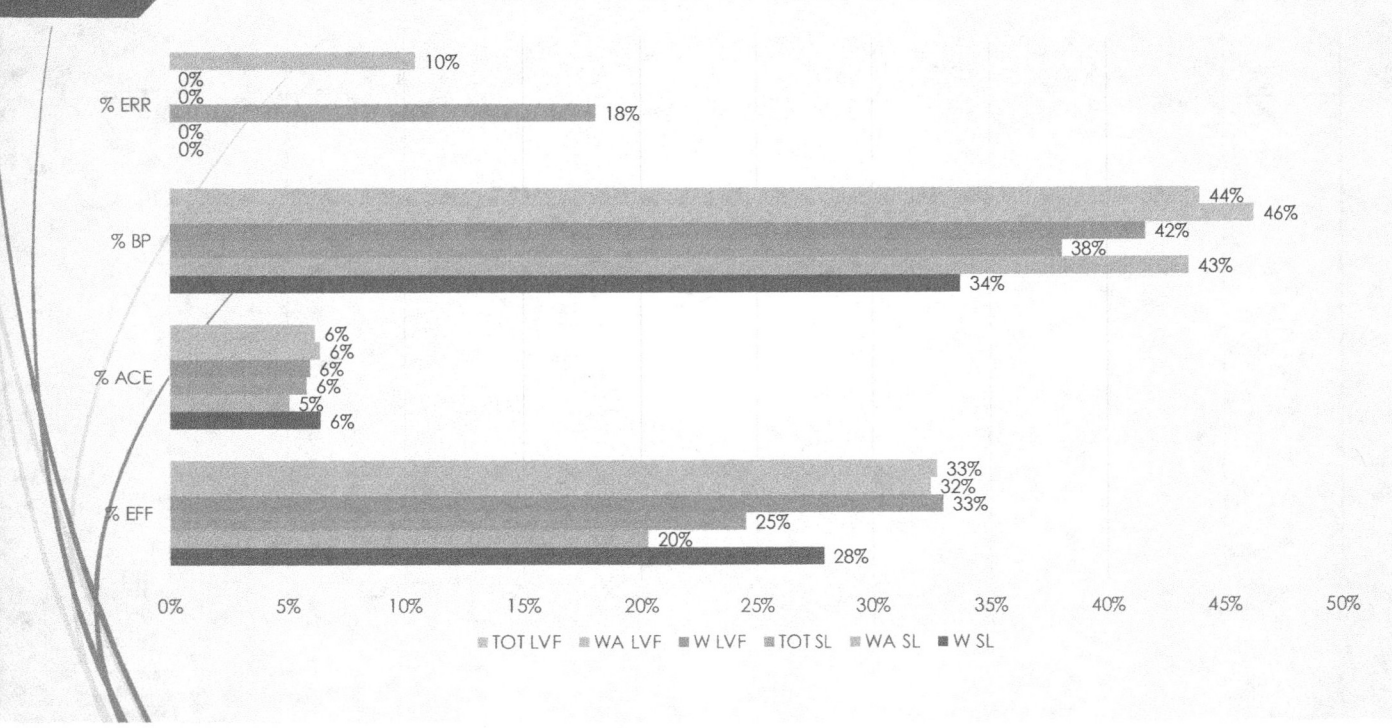

% ERR
- 10%
- 0%
- 0%
- 18%
- 0%
- 0%

% BP
- 44%
- 46%
- 42%
- 38%
- 43%
- 34%

% ACE
- 6%
- 6%
- 6%
- 6%
- 5%
- 6%

% EFF
- 33%
- 32%
- 33%
- 25%
- 20%
- 28%

0% 5% 10% 15% 20% 25% 30% 35% 40% 45% 50%

■ TOT LVF ■ WA LVF ■ W LVF ■ TOT SL ■ WA SL ■ W SL

FALLING BALL AND DROP: THE FUTURE

➡ THE JOKEW(DROP IN ZONE 4-5) AND WA (IN ZONE 1-2) ARE BATCHES THAT HAVE A HIGH % EFFICIENCY, BP AND ACE COMPARED TO THE OTHER TARGETS.

➡ THE WA, IN ZONE 2-1, DUE TO THE EASE OF IMPRISING ROTATION AND MASKING THE POWERFUL BALL, HAD AN EXTENDED EFFICIENCY % OF 33% AND 25%, AND A BP EVEN OF 45%, CERTAINLY THE HIGHEST BETWEEN THE JOKES.

➡ COMPARING THE % OF ERROR WITH OTHER TYPES OF KICKS IT IS CERTAINLY A SERVICE THAT IS HIGHLY EFFICIENT WITH RESPECT TO COST, BUT IT NEEDS TRAINING

➡ PERSONALLY I BELIEVE THAT A PLAYER MUST ALSO POSSESS THESE TECHNIQUES, RECOGNIZE THEM AND KNOW HOW TO ADOPT WHEN THE TECHNICAL OR TACTICAL SITUATION REQUIRES HIM INSTEAD OF HAVING HIMSELF INTO THE KNOWN AND SAFE.

➡ TO BE SATISFIED WITH THROWING IT THERE CANNOT BE THE PROFESSIONAL ROUTE FOR A SPORT THAT IS SO TECHNICAL AND IN WHICH THE BIT HAS SO MUCH AFFECT ON THE EFFECTS OF THE END POINT.

IN CONCLUSION

- OFTEN THE JUMP SPIN IS A MERE POWER EXERCISE WITHOUT PRECISION.
- HOWEVER MOST ACE ARE THE FRUIT OF PRECISION (TARGET G AND B) AND VARIATIONS (TARGET W AND WA).
- CERTAINLY WHAT IS MISSING IS THE COURAGE TO TAKE RISKS (BOTH TECHNICIANS AND PLAYERS), BOTH IN TERMS OF ANGLES AND IN TERMS OF VARIATIONS, OR TO EXPLORE LITTLE KNOWN AND AT LEAST IN EFFICIENT LANDS.

ALWAYS REMEMBERING THAT THE DIFFICULTIES THAT TEAMS ENCOUNTER, THE RECEPTION LINES, THE INDIVIDUAL RECEPTORS, DIFFER FROM PERSON TO PERSON, AND IS THEREFORE FUNDAMENTAL AN ANALYST OF PASSING FOR:

- **RECOGNIZE WHAT GIVES YOU DIFFICULTY**
- **KNOWING HOW TO REPRODUCE IT WITH A VALID SERVICE IN TERMS OF PRECISION AND SPEED**

Analysis of SERVE speeds between Super League and LVF 2021-22

▶ Pulling harder isn't always the best solution for everyone.

Ace for speed Jump Spin Super League

- ace slash efficiency mistakes
- Js below 100 km/h 46 out of 1085 0.42% 38 3.5% 175 16% -12.1%
- Js drop > 66 kmh 16 out of 459 3.4% 20 4.35% 42 9.1%-1.3%
- Js between 100 and 110 km/h 43 out of 818 0.52% 33 4% 211 25.7%-21.2%
- Js between 111 and 117 km/h 80 out of 732 10.9% 52 7.1% 180 24.5% -6.5%
- Js above 117kmh 36 out of 162 22% 13 0.9% 42 25.9% -3%

If your player doesn't serve above 100 kmh forget it.

Ace for speed Jump Float Super League

- ace slash mistakes eff
- Jf below 60 kmh 7 out of 330 2.12% 12 3.6% 12 3.6% +2.1%
- Jf drop < 50 kmh 2 out of 80 2.5% 4 5% 1 1.25%+6.2%
- Jf between 60 and 69 km/h 7 out of 319 2.19% 11 3.4% 26 8.1%-5.5%
- Jf between 70 and 75 km/h 3 out of 27 11.1% 0 0% 4 14.8%-3.7%
- Jf above 75 kmh 0 out of 20 0% 1 5% 8 40%-35%

If your player doesn't have power to shoot a jump spin above 100kmh you better need one jump float under 50km drop or even a jump float so long as slowly, under 60kmh.
However, if he has the power to go further, don't let him do a float serve above 75kmh because he would make more mistakes than throwing it jump spin above 110.

Ace for speedJump Spin LVF

- ace slash mistakes eff
- Js total kmh 84 out of 1156 7.2% 77 6.6% 170 14.7% -0.3%
- Js drop <50 km/h4 out of 1921% 0 0% 9 47.3% -20.5%
- Js drop 51><67 kmh 2 out of 38 5.2% 1 2.5% 6 15.7%5.5%
- Js between 85 and 95 km/h 31 out of 578 5.3% 36 6.2% 90 15.5%-4%
- Js between 96 and 100 km/h 31 out of 245 12.6% 18 7.3% 43 17.5% +2.4%
- Js above 100 kmh 1 in 8 12.5% 3 37.5% 2 25%+25%

If your player doesn't have the strength to pull a Jump Spin above 96 km/h it's best to leave it alone.
No one knows how to do the women's drop at 44kmh, but many do it below 66kmh but without benefits although still with a better efficiency than the spin up to 95kmh.

Ace for speedJfLVF

- ace slash erroreff
- Jf total 119 out of 3063 3.8% 191 6.2% 321 10.4% -0.4%
- Jf under 47><60 kmh 37 out of 1290 2.8% 74 5.7% 105 8.1% +0.4%
- Jf drop < 47 kmh 5 out of 157 9.5% 9 5.7% 17 10.8%+4.4%
- Jf between 60 and 69 km/h 61 out of 1531 3.9% 92 5.9% 162 10.5%-0.7%
- Jf between 70 and 75 km/h 13 out of 122 10.6% 16 13.1% 25 11.2+12.5%
- Jf above 75 km/h 8 out of 120 6.6% 9 7.5% 29 24.1% -10%
- Jf above 81 km/h 3 out of 77 3% 5 6.4% 21 27.2% 17.8%

If your player can't/can't shoot a jump float above 70kmh, it makes little difference whether it pulls slowly or strongly up to 69kmh, at which point a jump float under 46km drop that falls. However, if he has the power to exceed 70 kmh it is a good idea to let him do a float above 70kmh, but not above 75kmh because he would make more mistakes without great ace advantages.

Males vs females Speed Analysis

- For both men and women it is better to go for power if you have hitters capable of serving Jump Spins above 117 and in any case even above 110.
- If you don't have players with power though:
  - for men, a Jump hitter is better Foat with a joke that falls,
  - for women a Jump Float line between 70 and 75 km/h is better.
- For both genders the serve over 75 km/h terribly raises the % of error so it is not to be sought as a working objective, while for both the ball that is between 70 and 75 km/h is the ball with efficiency greater.
- For everyone, the falling drop is a winner but women don't know how to do it.

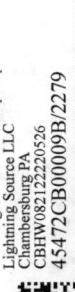